INVENTOR'S GUIDE TO IDENTIFYING PROFITABLE IDEAS

By: Jon W. Mooney

Edited with foreword by: John L. Janning

Published by:
Rapid Product Development Press
Amelia, OH

By: Jon W. Mooney

Edited with foreword by: John L. Janning

Published by: Rapid Product Development *Press*
P.O. Box 655
Amelia, OH 45102-0655

ISBN: 0-9653592-0-4 17.95

Library of Congress Catalog Number: 96-92512

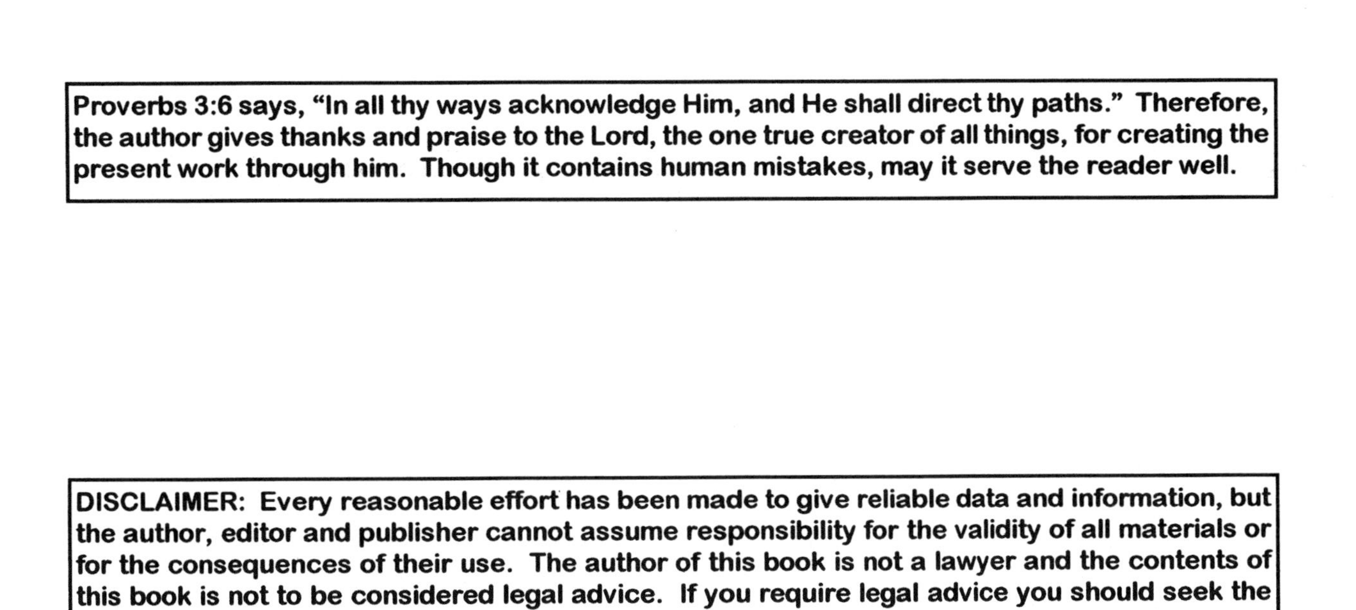

Proverbs 3:6 says, "In all thy ways acknowledge Him, and He shall direct thy paths." Therefore, the author gives thanks and praise to the Lord, the one true creator of all things, for creating the present work through him. Though it contains human mistakes, may it serve the reader well.

ADDITIONAL COPIES of this book may be purchased directly from the publisher by sending a check or money order in the amount of $19.95 for each copy requested. A 20% discount applies to orders of five or more copies.

Send request and remittance to:

"Inventor's Guide to Identifying Profitable Ideas"
Rapid Product Development Press
P.O. Box 655
Amelia, Ohio 45102-0655

For Lynn,
Joshua,
Thomas, Sarah
and James

ACKNOWLEDGEMENTS:

Thank you to all who inspired and supported the writing of this book. Thank you to all of my clients, past and present, whose common need for professional guidance at an affordable cost was the inspiration for this book. Thank you to John Janning for taking time out of your busy schedule to critique and edit the manuscript and for writing the foreword. Thank you to my family and friends for your love and moral support.

On the Cover:

"Brainstorm", Copyright 1996 by Ed Ostendorf, Jr.
&
Mimi Ostendorf Smith
(used with permission)

Illustrations:

Cartoons by Matthew R. Mooney
of
Grey Wolf Studios

TABLE OF CONTENTS

FOREWORD

Ideas - the most misused, misunderstood, misconcepted and ubiquitous word in the english language. Ideas are part of each and every one of us and common to our everyday living like eating and breathing. Even the birds and bees, along with all the animals and insects, get ideas. They may not be on the same plane as humans but put some bread crumbs in the yard and watch the birds get ideas. They get the idea that this might be something good to eat. After a while, the yard is full of birds. Oh sure, you can pass it off as instinct. But so what? At least they are doing something that is constructive with what was put in front of them. Even an ant knows how to build a house and does.

We, as humans, have a lot more going for us than the birds and the bees. We can take what is in front of us and create something that has never been before. We have the ability to change the world - at least our own world. Many do just that. But the majority do not understand the meaning of *ideas*. They think that ideas are all that one needs to make a lot of money. The biggest misconception of all time is that ideas are valuable. Most think that just because they have a good idea, everybody else should support them or listen to them. They look for someone else to fund their project because they can not afford to do it alone. Or they look for someone to *buy* their idea and put it into production so they can sit back and just cash the checks. Well, that is how it works in Utopia but not on earth.

On earth, Nicholas Murray Butler tells us there are three kinds of people. There are a few that make things happen; there is a larger number that watch things happen; but the overwhelming majority just do not know what is happening. Paraphrasing this another way, there are a few who diligently pursue their ideas by making prototype models - one after the other - without constantly thinking about how much money they are going to make from it but rather how nice this will be or how helpful it will be for their fellow man. Then there is a larger number who work on their ideas for a little while but get frustrated when they run into a problem. They do not want to invest much time, money or sweat to make model after model. They would just like to 'cash-in' as soon as possible. Now we come to the third group - the overwhelming majority - who just have ideas. They do not want to do any work; invest any time or money; search the prior art, etc., etc. They just want to sell their ideas. Somewhere along life's path, they ran into the greatest misconception of all time - that ideas are all that is necessary to make money. For some strange reason (unknown to science), they do not understand that getting ideas happens to everyone. Many believe that *their* ideas are better or more numerous than others. These are the most frustrated group. They can not understand why no one will buy their idea; listen to their idea; or fund development of their idea.

Ideas are worthless. Inventions are valuable. The difference between an idea and an invention lies in the reduction to practice. Once you make a model of your conception, it becomes invention. While the idea is just in your head, it is simply still an idea.

The path to riches is full of sweat, disappointments, failures, sacrifice, late hours and hard work. There is no easy way. If one accepts these heartaches, the chance of success gets closer. Jon Mooney describes the necessary steps in this book. He gives you names and addresses of valuable contacts. He teaches how to evaluate your idea and guides you to think like the professional.

Good luck on your journey to the world of inventing. Keep your eye on the road and follow these three signs: 1.) Forget about all the money you might make - just do a good job at making your prototype. It does not have to be expensive but needs to show functionality; 2.) Keep a good notebook of your activities, dates, witnesses, etc. 3.) Follow the guidelines in this book. It might be wise to read it through twice so nothing is missed.

Success and fortune follow those who perservere - whether they seek it or detest it.

John L. Janning

Inventor - LCD Alignment; Thermal Printing; Plasma Displays
34 U.S. Patents + >200 worldwide

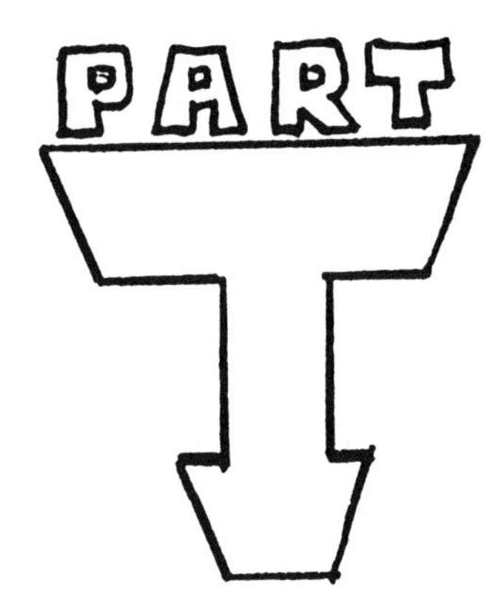

Identifying Profitable Ideas through Product Evaluation

SEEING NO OTHER ALTERNATIVE, RALPH DECIDES TO DIVE HEADFIRST INTO HIS LATEST VENTURE

Chapter One - "Look before you leap (then take the elevator)." - STANLEY MARTT, 1972

Every year, millions of dollars are wasted in the development, patenting and marketing of products which have no chance of making a profit for their investors. The goal of this book is to give the product developer and the inventor of new products the means to identify and discard over 90% of these useless products before precious development money and time is invested in them.

New product development and inventing is characterized by many false starts and dead ends. Logically, a creative product development group will come up with many more product ideas than the group has resources to develop. From this long list of ideas, management is usually expected to choose the best one to invest in. For the independent inventor, one product developed over the course of a lifetime is not uncommon. Unfortunately, the decision of which idea to pursue is often one based on one individual's favorite or "pet" idea. Without an unbiased and standard method of evaluating the potential success of a new product idea, neither management nor inventor is capable of consistently "separating the grain from the chaff."

How many new product ideas eventually become successful products? A visit to a Patent Library will give you some idea. The United States Patent and Trademark Office (USPTO) currently issues about 100,000 patents each year. Look through the Patent Gazettes and you may find one product that you recognize as successful for every thousand that you have never even heard of before. Assuming this success rate is the same for unpatented product ideas also, the chances that any new product idea will become profitable is about 1 in 1000.

Since no company, and certainly no one inventor, can afford to develop 1000 products just to see one succeed, effective and serious evaluation must be the first job of the successful product developer. The goal of successful product development then must be to spend the least resources to weed out the products which have little chance for success. This frees up the majority of the development budget for those ideas with the best possibilities.

This evaluation evolved from the author's 17 years of experience as a product development engineer and from his own business as a consultant to independent inventors. It is the author's experience that 9 out of 10 ideas evaluated using this method are rejected. Obviously, this ratio will vary both with the quality of the research and the quality of the ideas being evaluated.

TOYS AND GAMES

A word about Toys and Games: the Toy Industry is one of the most competitive and therefore, it is very difficult for a nonmember to penetrate. The success of a product here depends heavily on television advertising budgets. If your idea is one for a toy or game, and you are not already a member of the toy industry, consider taking the idea to a Toy Agent. Toy Agents can evaluate your idea for a minimal charge and, if they like it, will attempt to sell the idea to their Toy Manufacturing clients. If successful, it will cost you a high percentage of your royalties, but this may be your only way to break into the toy industry.

Toy Agents can be found in the Annual Buyer's Guide of Playthings. You may be able to find this magazine at your library. If not, you can write the publisher at the following address:

Geyer-McAllister Publications, Inc.
51 Madison Avenue
New York, NY 10010

You also may be interested in the "Guide for Toy Inventor/Designers"; a free publication of the Toy Manufacturers of America. Their address is:

Toy Manufacturers of America, Inc.
200 Fifth Avenue, Suite 740
New York, NY 10010

INDEPENDENT PRODUCT DEVELOPERS

If you do not have the time to perform your own evaluation, then you may want to consider employing the services of an Independent Product Developer (IPD). As an IPD, I know that this field is a magnet for scam artists in search of a quick buck; so beware. One way to determine if an IPD is legitimate is to compare copies of previous evaluations for several products. Most scam artists will use the same evaluation form for every product and simply "fill in the blanks." You can literally hold two different evaluations together up to a light and most of the words will line up. Also, the scam evaluation will not give market information which is specific to the product being evaluated. Instead, the scam evaluation will show figures for an overall market.

Keep in mind that the number of successful products which an IPD has developed is only an indication of the quality of new product ideas brought to him and how long the IPD has been in business and therefore, is not a good measure of the IPD's capabilities. If you ask an IPD how many of his clients' products have become successful and the answer is any more than 1 in 1000 —— BEWARE! Either the IPD is not counting some clients or is redefining the term - "successful." Independent Product Developers cannot make a product successful if the product idea does not have the potential for success. A product becomes successful only if it is the one in a thousand that has the potential for success. You cannot buy success!

USELESS PATENTS

Many inventors rush directly to a patent attorney when they are struck with a new idea. But no one would think to ask a patent lawyer how many of his clients' products have become successful. It is the author's opinion that the number of successful products would be a good measure of the patent system's economic value — since each patent now costs its inventor over $6000! Just consider the money wasted each year pursuing useless patents: 100,000 patents issued each year at an average cost of $6000 per patent with a success rate of only 1 in 1000 equals $599 MILLION wasted each year! If we could weed out just 90% of these wasteful patents, the $540 million saved could be intelligently used to develop and patent the ideas having the best chances for success and to create new jobs manufacturing those products.

$540 MILLION DOLLARS A YEAR. IS SPENT ON WASTEFUL PATENTS.

OTHER SOURCES FOR PRODUCT EVALUATIONS

Another source for product evaluations is the Wal-Mart Innovation Network (WIN). For a small fee, the students at Southwest Missouri State University will evaluate and score your product idea for potential. If the score is high enough, and Wal-Mart buyers are interested, they have been known to place orders with the inventor to supply their stores with the product. To find out more about this program write to the following address:

Wal-Mart Innovation Network
Center for Business and Economic Development
Southwest Missouri State University
901 S. National Avenue, Springfield, MO 65804

Call your local College of Business Administration to check if they have a product evaluation program similar to Southwest Missouri State's.

One last method of evaluating potential success is to test market the actual product. If a small quantity (say 100) of the product can be inexpensively made and placed on the store shelves next to competing products, your rate of sales compared with those of competing products will tell you exactly what the potential success of the new product might be. If you do this, be certain that the price of your product is close to your estimate of what the item would be in large production. Since products are generally more expensive to make in small quantities, this may require you to price the product so that you actually lose money on the test market items. Consider this "loss" as an investment in real-life market information. Also, you should be aware that if you intend to pursue a patent on the product, this must be done within a year of the first public disclosure or sale of the product.

Chapter Two - Instructions

The evaluation is a list of 20 questions to which you must answer "yes", " no" or "maybe". Each "yes" answer is scored 5 points. Each "no" answer is scored 0 points. Each "maybe" answer is scored 1 through 4 depending on the surety of the answer. You will need to use your judgment in scoring some of the questions but do not get too concerned about any one answer — score a 3 for any that you are not sure about. The evaluation is not more detailed since the information you will be using is not that specific and since you are only looking for a yes or no result on whether to proceed with development. After answering all of the questions, you simply add to find a total score. Just like in school, 95 to 100 is an "A", 85 to 95 is a "B", 75 to 85 is a "C", 70 to 75 is a "D" and below 70 is an "F". Depending on how much risk you wish to take, you now can use this score to decide whether to continue with the development. When I evaluate a product, it must rate an "A" or "B" for me to recommend development.

It is always tempting to "save time" and evaluate an idea without gathering the necessary information. You may think that you already know the answers to the questions on the evaluation, especially if they are 5's. It has been my experience, however, that wishful thinking blinds even the best of us, and that filling out the evaluation without first gathering the information required is a rationalization and a waste of time.

LIMITATIONS

This evaluation does have its limitations. Any product whose success is driven by huge amounts of advertising directed towards "fad" markets (for example toys, games and pet rocks) will not score well in this evaluation. The evaluation works best for stable markets with well-defined customers. The accuracy of the evaluation will improve with your diligence in gathering the proper information, honesty in answering the questions and overall experience.

The evaluation concentrates on the marketability of the product, the market for which the product is designed and the manufacturability of the product. This should be sufficient for the independent product developer and inventor. If you are evaluating a new product idea for a company, you must also consider how well the product fits in with the manufacturing capabilities, financial requirements and target markets of the particular company.

HOW TO START

The Product Idea Evaluation is given on the next two pages. It is given here for you to review and to direct your subsequent search for information. Read through the evaluation sheet now and memorize the type of information you are looking to find. As you review the information sources which are given in the next section, remember the questions and try to find their answers. Sorry, but there is no one resource that I can direct you to where you will be certain to find any particular piece of information in your search. You will need to keep on your toes, read, understand and apply the information you find. Just as in any successful hunt, you must constantly keep in mind what it is you are looking to find.

YOU MUST CONSTANTLY KEEP IN MIND WHAT IT IS YOU'RE LOOKING TO FIND.

PRODUCT IDEA EVALUATION

PRODUCT IDEA:____

QUESTION	SCORE (1 - 5)
1) Is the market size large enough?	________
2) Can the market be reached easily?	________
3) Can you expect a large enough share of the market?	________
4) Are there several markets for the product?	________
5) Will the industry continue to grow?	________
6) Is the market steady?	________
7) Will the product revolutionize the industry?	________
8) Is the product's price lower than the competition's?	________
9) Is the product's performance better than that of the competition?	________
10) Are there established distribution channels that the product can use?	________

PRODUCT IDEA EVALUATION

PRODUCT IDEA:____

QUESTION	SCORE (1 - 5)
11) Is the product very different from the competition?	________
12) Are there only a few big competitors?	________
13) Will the demand for the product last several years?	________
14) Will the product be accepted by the public and consumer organizations?	________
15) Is the product technically feasible?	________
16) Can the design problems be solved in a short time?	________
17) Does the product use readily available materials?	________
18) Are only simple manufacturing and assembly operations needed ?	________
19) Is the product patentable?	________
20) Are there few product regulations in this industry?	________
Total Score:	________

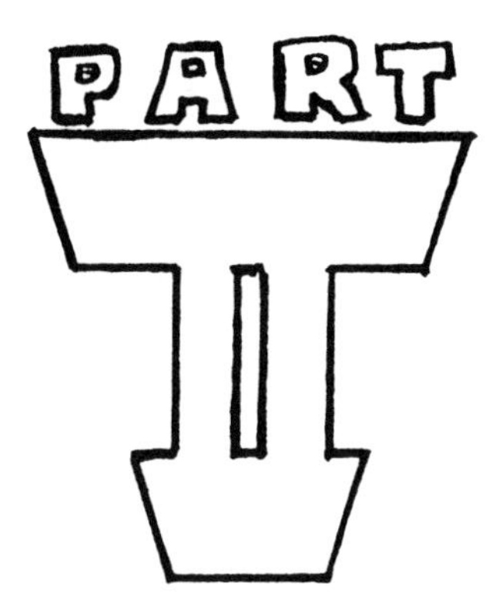

Gathering Information for the Evaluation

MAN GUYS, IF THAT'S THE COMPETITION, THIS IS GOING TO BE EASY!
WATER BOY
WA
ULP!

Chapter Three - The Search for Competitive Products

Your goal in this section is to find all of the products which are competitive to the new product idea. These competing products may be similar in appearance to the new product concept or very different in appearance but have the same purpose.

It is unlikely that the new product idea will have no competitive products. The new product concept is a proposed solution to some problem. The problem may be one of great public concern or just what to do in leisure time. Your intended customers are presently using competing products to solve this problem. Find out what these products are, who makes them and how many are sold each year.

The existence of competing products is a good sign of the marketability of a new product. The lack of competing products is an indication that either no market exists for the new product concept or that a large advertising budget will be needed to educate the public about the new product and to develop a market for the product.

As you gather information, fill out a "Competitive Product Worksheet" (see Appendix A) for each competing product.

BUYER'S DIRECTORIES

Purchasing agents use Buyer's Directories to locate companies that manufacture and distribute products and services for which they have a need. Most public libraries have sets of these. A common one is the Thomas Register, Thomas Publishing Co., NY. These directories list companies by the products that they manufacture and may even include a short catalog for some companies. Look under the headings which relate to your product to locate all of the companies which manufacture competing products. Call or write each company asking for company information, sales brochures, catalogs and product specifications for their line of products. Information about company size and sales is as important as the information on the products and pricing. Keep information on competing manufacturers in Appendix B.

CATALOGS

The target market(s) for the new product idea may be one for which mail order catalogs exist. Browsing through these catalogs will not only reveal competing products but will also show you what other products your target market consumer buys. This information will be important in defining your market. Good sources for locating catalog companies include The Directory of Mail Order Catalogs, Cottlieb,Grey House Publishing, CT, 1996, The Catalog of Catalogs, Palder, Woodbine House, MD, 1995, and The National Directory of Catalogs, Oxbridge Communications,Inc., NY, 1995.

MAGAZINES

Chances are good that your target market is served by one or more specific magazines. Ask your librarian for a reference book of periodicals. Two popular ones are Ulrich's International Periodicals Directory, NY, Bonker, and The National Directory of Magazines, NY, NY Oxbridge Communications. Identify the magazines which serve the market you are interested in and list them on the "Periodical Worksheet" (see Appendix D). If the library has back issues of these magazines, browse through them to identify advertisements for similar products as yours. Write or call these competitors to obtain product and company data.

Past issues of the magazines that you have identified may carry articles on competing products. You can search for this information by using the annual periodical subject indexes and abstracts. Some useful ones are listed below.

Access: The Supplementary Index to Periodicals
John Gordon Burke Publisher, Inc., US

Alternative Press Index
Alternative Press Center, Inc., MD

Applied Science and Technology Index
H.W. Wilson, NY

ASCE Annual Combined Index
American Society of Civil Engineers, NY

General Science Index
H. W. Wilson Company, NY

Index to IEEE Publications
Institute of Electrical and Electronics Engineers, Inc, NY

Names in the News
NewsBank, Inc., CT

Reader's Guide to Periodical Literature
H. W. Wilson Company, NY, Cambridge, London

The Engineering Index Monthly
Engineering Information, Inc., NJ

THE EFFICIENT PATENT SEARCH

You can perform a preliminary Patent Search which, although probably not thorough enough for a patent application, should allow you to determine if your exact idea has been patented by someone else. It is the author's experience that 7 out of 10 product ideas will have already been patented.

The existence of a previous patent on a product idea can actually be a good thing. If you do find a patent for the product idea you are researching, call or write to the inventor or patent attorney and ask the following questions:

1) Is the patent still in force?
2) Is the patent licensed to someone?
3) Is the product covered by the patent being manufactured?
4) If the product is not being manufactured, why not?
5) Does the inventor have other similar patents?

"WHAT WAS THAT ?!!"

To keep a patent in force, the inventor must pay maintenance fees to the Patent and Trademark Office at 3 ½, 7 ½, and 11 ½ years after the patent is issued. If these fees are not paid on time, the patent will normally expire. If the patent which you have found is expired, then it no longer prevents you from making or selling the product.

The existing patent may be licensed to someone. If this is true, then you need to find out if the license is exclusive or not. An exclusive license gives its holder the right to manufacture and sell the product in all markets and regions and would prevent you from manufacturing and selling the product. If however, the license is only for a particular market or region, you may want to consider purchasing a license from the patent holder to give you the right to manufacture and sell the product in another market or region.

If the product covered by the patent is being manufactured, find out who is manufacturing it, who is buying it, what it costs and how many are sold each year. If the product is being manufactured and sold to the same market that you have identified, then you may not want to pursue the product any further. If the product is not being sold to the market that you have identified, then you may want to consider a relationship with the manufacturer to distribute the product to your chosen market.

If the product covered by the existing patent is not being manufactured, find out why not. The inventor may have spent all of his money getting the patent and is just waiting for someone wanting to license it. More likely, the inventor will have found out too late, that the market for the product was too small, or the product was too expensive to make. If this is the case, then the inventor's misfortune will be invaluable market information for you.

Let me say this again. The existence of a previous patent on a product idea can actually be a good thing. If you decide to manufacture and sell a product, the existence of a previous patent which is available for licensing greatly simplifies your situation. The cost of a license is negotiated - the cost of a patent is an unknown. The cost of a license can be amortized over the manufacturing life of the product - the cost of a patent is paid up front. Your protection under a license is immediate and certain. Patents typically take 1 to 5 years to obtain and may be rejected after investing in their application. There are product developers and inventors who will quit the pursuit of a product if they can not have their name on a patent, but this is a decision made on vanity, not on logic.

HOW TO PERFORM AN EFFICIENT PATENT SEARCH

Patent searches are done at libraries which have been designated as Patent and Trademark Depository Libraries. There are currently 78 such libraries in the country. A list of these is included in Appendix E. It is the author's opinion that patent searches for the purpose of evaluating new product ideas are best done by the inventor or product developer. A professional patent agent performs a search with a mindset of locating any patents which the product may illegally infringe upon. A patent which does not meet this specific criteria is usually not copied or considered, even if the patent covers a competing product. A patent search done by the inventor or product developer will consider every competing concept without limitation. Although a patent agent should be employed when applying for a patent, a patent search for the purpose of this evaluation must be done through the eyes of the person most familiar with the concept. Otherwise, too much important information will be dismissed.

A word of caution: With nearly six million patents in existence and over 100,000 new patents issued each year, you could very well spend the rest of your life doing the patent search on just one product concept. Place a time limit on your search, before starting. Give yourself a week to learn how to use the system the first time. After you have researched several product ideas - you should average about 4 hours for an adequate search - there is a likelihood of finding an existing patent 7 out of 10 times.

Use Appendix F, "Patent Search Worksheet", to help in your patent search. The patent search starts with the Index to the US Patent Classification System. This is a subject index of patent classification numbers. Using the Index, look up every word that describes or relates to the product that you are researching. Be very broad-thinking when you do this; a concept for a police baton may be covered by a patent for a new type of baseball bat Next to each subject that you find listed will be a series of numbers and/or letters.

These are the class and subclass designations for the chosen subject. Write these down along with the subject word.

Next, locate the Manual of Classification. This is usually a three volume looseleaf binder. Look up the classification numbers that you found in the above Manual. The Manual breaks up the subjects into more precise classifications. Look down the list of subclass definitions to find all of the ones which categorize your product concept more precisely. Write these subclass numbers down along with their definitions. If you want, you can find better definitions of these subclasses in the Class and Subclass Definitions manual.

Now you are ready to use the CD-ROM database called CASSIS. CASSIS is an acronym for "Classification And Search Support Information System." Most Patent Libraries will have this database. If yours does not, you can still search for existing patents using the annual indexes, but your search will take much longer. Load the CD-ROM titled "BIBLIOGRAPHIC/ASSIGNEE FILES". After the database is loaded, you will see a screen which allows you to search for patents using any of several different methods. You can search by title or abstract, classification, patent number, issue date, State/Country, status, USAPatent Vol ID, assignee name or assignee code. You will be searching by classification.

Enter a classification number from your list and press enter. If you want to enter a class and subclass, type the classification number first, followed by a "/" and then the subclass number. The program will search the database and come back with the number of patents which match your selection. Assuming this number is greater than zero, select the function key which displays the results (presently F3).

The program will read the database again and come back with a screen listing the titles of the most recent patents which are in your selected classification. Read through these titles and note the ones which sound similar or competitive to your product concept. You can use the ⇧ and ⇩ keys along with the "page up" and "page down" keys to go through the entire list. Stopping on one title, you can view a small abstract of the patent by pressing "enter." To go back to the list, press "enter" again.

You can print out the list of patents by selecting the output function key (presently F4), and choosing "print." The list can also be saved to a floppy disk using this function.

Write down every patent number which appears to be closely related to your product concept or print out the entire list if desired. Do this for every class and subclass number which you have identified.

The majority of your patent search will be spent in the next task by reviewing the abstracts of the patents that you have identified as possible competitors. This is done by looking up each patent number in the appropriate Patent Gazette. The USPTO publishes at least one Patent Gazette per month. In particularly creative months, they may publish two or three. These gazettes contain abstracts of every patent and trademark which is issued during the month. Patents are listed numerically by patent number but arranged by the classification number; trademarks are arranged by date of issue. The gazettes also contain lists of patents which have expired, lists of patentees and several other useful indices.

The gazettes will most likely take up one whole wall of your library. Using your list of patents from the previous step, look up the most recent patent on the list in the appropriate gazette. Note that the range of patent numbers covered is printed on the spine of each volume, along with the volume number and date of the gazette. To make your search efficient, find the right book by referring to the patent numbers and reshelve the book by referring to the volume number.

An abstract will usually include an illustration which will help you decide quickly if the patent is of interest or not. I suggest that you photocopy any abstract in which you are interested, there will be too many of them for you to remember which ones were related to your idea. Also, be sure to look at the patents nearest the one you are looking up. These will be in similar classifications to your idea. It has been this author's experience that, more than once, a patent for the idea at hand was found using this method.

When reviewing the abstracts, do not limit your interest to only those patents which work exactly the way that your concept does, as you should be interested in every concept which directly competes with your concept, no matter how it works. This is why a patent search done by the inventor is much more valuable than one done by a patent agent. Locating competing products simplifies your search for directly applicable marketing information and allows you to make informed decisions concerning the value of investing time and money on your concept. A search by a patent agent will, at best, only tell you whether the agent found patents which your idea may infringe upon.

Review the abstracts for each patent number on your list, making photocopies of the ones which are most interesting. Using these photocopies as a list, your next step is to review the full patents for these inventions. You will find the full patents in your local patent library, either on microfilm or in bound texts of patents.

Patents tend to be long-winded legal documents which are difficult to read or understand, but do not let this discourage you. The information that you are interested in at this point can probably be found on the first two and the last page. Photocopy anything that interests you.

On the first page of the patent, you can find the name and address of the inventor and the name and address of the patent attorney. If the patent was assigned to a company or third party, their name and address will also be found on the first page as the assignee. If the patent is for a concept exactly like yours or competitive to yours, copy these addresses so that you can later contact these people for market information. Keep information on each competitive patent on a copy of Appendix G, "Competitive Patent Information Sheet."

Also, on the front page will be a list of patents which the primary examiner referred to during the USPTO patent search. Copy these numbers down so that you can later go back and review their abstracts in the Patent Gazettes. This may seem like a lot of tedious work but a thorough review of the patents can reveal valuable marketing information which will greatly save your time and money in the long run.

An abstract which is similar to that found in the gazette will be included on the front page of the patent. After this abstract, the inventor will usually write a summary of the history and state of the art of the invention. This section may contain information on your market, including other competing products and problems which have been encountered with different solutions. Also, the inventor may include the patent numbers of related inventions which you should make note of for later review in the gazettes.

Except for reviewing the illustrations, you will probably want to skip the body of the patent and go to the last section which contain the claims of the patent. The claims are the descriptions of what the patent protects and is the heart of the patent. Every patent has one or more claims. Design patents (prefixed with "D") have only one claim which simply protects how the invention looks, not how it functions. Utility patents have one or more claims which protect how the invention functions. The first claim is the broadest claim and protects the most general functions of the invention. The second claim is more specific than the first and therefore is less encompassing. The more descriptive and specific a claim is, the less protection it affords the inventor, since it is easier to design a competing product which does not infringe upon a more specific claim.

Review the claims of inventions similar to yours to determine if your product can be successfully designed so as not to infringe upon the patent. Make photocopies of any patent which contains claims which you must be concerned with when developing your new product. This is valuable product design information which will help in estimating the costs of developing your new product idea.

After reviewing all of the patents on your list, go back and review the abstracts for the additional patent numbers which you found referenced. This may take you to reviewing patents issued during the turn of the century, but do not think that just because a patent is old, your idea could not possibly be covered. It was discovered that devices for mounting whip antennas on vehicles were already covered by a patent related to buggy whips. The patent search continues until you have exhausted all of your leads or until you have reached your predetermined time limit.

If you have the name of a patent attorney from an interesting patent but no address, you can find it in the annual publication of Attorneys and Agents Registered to Practice before the USPTO, which your library should have.

Write or call inventors or their attorneys as discussed previously to obtain information on the status of patented inventions. A phone conversation with the actual inventor is probably the most valuable information source you will find. Generally, inventors are happy to talk to anyone about their inventions and may even offer to send you a sample of their product. They will have been involved with your market for several years and can offer valuable information and suggestions for you. You have spent a lot of time finding out who these inventors are so do not ignore this source of free information and help now!

INTERNET

With the recent explosion of activity on the Internet and World Wide Web, the individual now has the power to conduct world wide searches for market information using various search engines, conduct market surveys through the use of discussion groups and advertise the product to the world in "Hypermalls." Many texts are available to educate the reader on how to use the Internet so no attempt will be made to do so here.

SEARCH ENGINES

Search engines are programs which contain indices to the various locations on the World Wide Web. To use these programs, you simply call them up on your computer and give them a topic to search. There are currently ten popular search engines. These are listed in Appendix H. The engine will look for your topic in its indices and then give you a list of all the locations which contain your specific topic word or words. Along with each location is one or two lines of text from that location to help you decide if it interests you. If you find a location which does interest you, simply select it to jump to that location on the Web. Your search can take you around the world into any one of millions of computers. Be warned, that since there is so much information on the Web and since it is scattered all over the world, the fascination of the hunt can

"OKAY, I'M READY TO SURF THE INTERNET"

tempt you into wasting a lot of time searching while developing little useful information. Fight any urge to browse by being very specific in your search. If a search engine finds several hundred locations which contain your subject word, then your search is too broad. Narrow the list down by being more specific, adding descriptive words to your search.

The use of search engines can and frequently does yield good, useful market information but they must be used with the same type of restraint as in watching TV. If not, the search will be more of a waste of time than informative.

USEFUL LOCATIONS ON THE WORLD WIDE WEB

The following are a few locations on the World Wide Web which the author has found to be particularly useful when researching a new product idea. Since the Internet is constantly changing, the locations given may be different or no longer in existence.

LOCATION ONE -
UNITED STATES PATENT AND TRADEMARK OFFICE HOME PAGE

The home page of the USPTO is located at [http://www.uspto.gov/]. It contains information about patents and links to other sources of information. The USPTO publishes several general booklets on the subject of Patents and Trademarks which are also listed here.

LOCATION TWO - PATENT SEARCH PAGES

The USPTO along with two other organizations (CNIDR and MCNC) have made the bibliographic patent database available on the World Wide Web. Its location is [http://patents.cnidr.org:4242/access/access.html]. Although convenient, this database has limited use in a patent search since the database only covers patents from the previous 20 years. The first page of these patents containing the abstract can be viewed or downloaded. This first page also contains references to prior patents related to the invention.

LOCATION THREE - SWITCHBOARD™

This is a searchable database containing phonebook information (name, address and phone number) for all businessess and individuals in the United States. This can come in handy when trying to locate an inventor who has moved from the address given on a patent in which you are interested. The location is [http://www.switchboard.com/].

LOCATION FOUR - DUNN AND BRADSTREET

You can purchase the D&B report on a company and download it into your computer using Dunn & Bradstreet's internet services. This may be useful when trying to estimate the size of your intended market by reviewing competing companies' sales records. Find this service at [http://www.dbisna.com/].

LOCATION FIVE - SIC CODES PAGE

ITA software corporation presently has the Standard Industrial Classification (SIC) Codes available on the Internet. The government classifies different business activities according to a four digit code. Knowing this code for the type of product that you are contemplating helps you narrow your search when looking through databases for competitors. The SIC Codes are located at [http://www.wave.net/upg/immigration/sic_index.html]. SIC Codes can also be found in the Standard Industrial Classification Manual, OMB, 1987.

CD ROM / ELECTRONIC DATABASES

Your nearest large city public library or university probably has several databases which will be useful in your search for competing products and manufacturers. If you visit the library, these will be available on special computers with CD ROM drives. If the library has a computer server, then you may be able to use your computer and modem to search these databases from your home. Databases which may be available follow.

DATABASE ONE - LIBRARY CATALOG

A quick search through the library catalog may reveal books, audio and video cassettes which relate to the history and present state of the art of related products and your target market. Also use this to locate the publications referenced throughout this workbook.

DATABASE TWO - BUSINESS & MANAGEMENT INFORMATION ABSTRACTS UMI (ABI/INFORM)

This database contains abstracts of articles published since 1992 in 800 business publications. It presently holds 892,614 abstracts. It is useful in locating both competing products and manufacturers and to determine what the prime motivators of your particular market are.

DATABASE THREE - NEWSPAPER ABSTRACTS (UMI)

This database presently contains 3,691,518 abstracts of articles published in 30 newspapers since 1993. If your product is conceived to solve a problem which is of general public concern, then this database can be used to find supporting articles for estimating the prevalence of the problem.

DATABASE FOUR - PERIODICAL ABSTRACTS (UMI)

A database of 2,262,812 abstracts of articles from 1643 publications since 1993 is useful for locating magazines which serve your market as discussed under "MAGAZINES" and "MEDIA KITS."

DATABASE FIVE - HEALTH INDEX, INFOTRAC 2000 (IAC)

This database may prove to be useful if your product belongs in the medical market. It consists of 248,162 abstracts from 102 health publications published since 1991.

DATABASE SIX - GENERAL PERIODICALS INDEX, INFOTRAC 2000 (IAC)

This one contains 2,062,742 abstracts from 1490 general publications since 1991 and has the same use as the PERIODICAL ABSTRACTS (UMI) .

DATABASE SEVEN - BOOKS IN PRINT

There are about 10 million listings in this database of recently printed books. This will expand the search you started with the LIBRARY CATALOG.

PREHISTORIC INVENTOR OG IS DISTRAUGHT UPON CREATING ANOTHER SEEMINGLY USELESS INVENTION.

Chapter Four - The Search for Market Information

Using the resources given in this section, you will next want to better define your specific market and find the answers to the evaluation questions that deal with your market. The information that you are looking for can be grouped into ten categories. Keep these categories in mind as you continue your search and place the information that you find in the proper section of the "MARKET WORKSHEET" given as appendix I.

TEN MARKET INFORMATION CATEGORIES

CATEGORY ONE - CUSTOMER

Define the potential customer as specifically as you can (for instance "the customers are college educated women, aged 25-34 or over 45 who are married with 1 or 2 children and have a $30,000 household income"). Being precise allows you to start thinking like the customer and aids in the design of your product. It also allows you to better estimate the size of your target market, using census-type data.

CATEGORY TWO - LEAD USERS

Identify the customers who will use your product the most. Ideally, you will be able to identify these people as members of certain organizations or user groups. Lead users can be an important source of market information and will be the most informed on problems involved with using competing products.

CATEGORY THREE - PROBLEM

Your product concept is a solution to some problem that you have had or have thought about. Talk to some of the potential customers that you identify. Do not tell them about the product idea but ask them about the problem that your product is designed to solve. You may be surprised to find that your potential customer perceives the problem differently than you had imagined or they do not believe a problem exists at all. An idea for a machine that automatically puts jigsaw puzzles together, although ingenious, will be useless to the potential customer who enjoys sitting for hours putting puzzles together. Additionally, you may find that your potential customers have already found a simple solution and do not need your product. If they perceive the problem differently, it may guide you toward another new or improved invention.

CATEGORY FOUR - HOW THE PRODUCT SOLVES THE PROBLEM

Answer this question from the point of view of the potential customer. Describe how the customer will use the product and what benefits the customer will gain. Do not design the product yet and do not describe how the product works - you do not want to limit your options at this stage. Concentrate on the desired benefits of the product - not the physical design features.

CATEGORY FIVE - VALUE OF THE PRODUCT

The best source of information on the value of your product are your potential customers. Talk to some of them to find out, in general, what they would pay to solve the problem you have identified. Make sure you talk to customers that would make only limited use of your product as well as those who would make constant use of it. Obviously, those that will use it frequently will place a higher value on the product. Do not deceive yourself into thinking that all of them would put such a high price on it.

CATEGORY SIX - PRODUCT COMPARISON TO THE COMPETITION

Describe how competing products operate, what benefits they offer and what they cost. Make a list of the benefits and costs to compare the products with one another and yours. This list will be useful to determine what benefits and cost your product must offer to the customer in order to be competitive.

CATEGORY SEVEN - POTENTIAL CUSTOMER'S OPINIONS

What benefits and features of your product are most important to the customer? Using the list from the previous section, ask potential customers to rate these benefits according to their importance. Have them rate the competing products for each benefit. This will give you an idea of what combinations of features your product must have to be competitive. For those of you who are interested in such things, this procedure is a basic step in the latest Japanese fad of Quality Function Deployment (QFD).

CATEGORY EIGHT - SIZE OF THE MARKET(S)

One of the biggest mistakes that people make is overestimating the size of the market for a product. If, for example, your idea is for a new type of bumper sticker, you should not fool yourself into thinking that the market size is equal to the total number of vehicles on the road. A simple survey of cars in any parking lot will prove that the percentage of vehicles which display bumper stickers is but a small fraction of the total. There may be two or more well defined markets for your product. The estimate of market size is only as accurate as your definition of the customer.

CATEGORY NINE - DESIGN PROBLEMS

Even if your product idea is a relatively simple one, there are always problems of one type or another to solve before an acceptable design is conceived. Design problems can be broken down into three categories: 1) marketing problems, 2) engineering problems, 3) manufacturing problems. Marketing problems involve the features and benefits of the product and how good each feature must be in order to please the customer. Engineering problems are concerned with the efficient physical design of the parts making up the product and the reliability of the operation of the product. Manufacturing problems include those which make the product easy and inexpensive to build.

CATEGORY TEN - MARKET TRENDS

Try to determine the growth rate, sales volume and latest innovations for the industry and most reasons why the market goes up or down. Some of this requires deductive reasoning on your part but the results can guide you in making design trade-offs later on.

SOURCES OF MARKET INFORMATION

Use the following sources to locate information in each of the ten market information categories defined in the previous section and repeated in Appendix I. Organize your information under the proper category in the appendix for easy retrieval during the evaluation.

SOURCE ONE - MEDIA KITS

Media kits are information packets sent out by the media (eg magazines, radio, television) to prospective advertisers. The packet contains information on the style and format of the particular media and its advertising, rate cards and a market analysis of their specific audience. If your target market is the same as that of a certain media, then their market analysis will be directly applicable to your search. Since there are so many magazines with specific target markets, it is usually very easy to find one or more that match your product's target market.

Review your periodical list in Appendix D. To obtain a media kit, call the classified advertising sales office for each magazine. Ask for a media kit and, while you have them on the phone, ask them for estimates of how many subscribers the magazine has and what percentage of the total market they think that covers.

If your product does eventually get developed, you will want to consider using these magazines to advertise your product, therefore keep these media kits to later aid in planning an advertising campaign.

SOURCE TWO - ASSOCIATIONS

Associations are almost as numerous and diverse as magazines. Look through the Encyclopedia of Associations, Detroit, Gale Research Co. for organizations which serve your target market. Listings in this publication normally state the number of members and the purpose of the organization. If the association publishes a magazine for its members, this will also be listed telling if advertising is accepted.

Write to each applicable association and ask for information on membership and request a media kit for any of their publications. The information which you receive should contain detailed information about the organization and may be useful to you as information about your lead users.

SOURCE THREE - CUSTOMER SURVEY

If your target market is well defined, then the best source of information about the market is from the market itself. This is especially true when trying to find how the customer perceives the "problem", the value of the product, and the customers' opinions. The survey can be as simple as a conversation with several potential customers or as complicated as a focus group study. An excellent book on this subject is Measuring Customer Satisfaction: Development and Use of Questionaires, Hayes, ASQC Quality Press, Milwaukee, 1992. You may also adapt your competition's warranty questionaires for this purpose.

SOURCE FOUR - CONSUMER PRODUCT REVIEW MAGAZINES

The government tests and reports on the safety and function of various types of consumer products. A review of the annual index to Consumer Reports News Digest, Consumers Union of the United States, NY, is likely to reveal a report on products which compete with your product idea. Other markets in which products are frequently rated by industry magazines include consumer electronics, automotive and sporting equipment. Check the periodical abstracts as explained previously for published product reviews.

SOURCE FIVE - ANNUAL STATEMENTS

Public companies must file annual statements to their stock holders. These statements should be available in the Government and Business section of your library. If companies listed on your competitive manufacturers list are publically held, then their annual statement should be a good source of information. Annual reports often contain detailed information on target markets. The size of the market may also be estimated from the company's annual sales figures.

The Evaluation

"AND, THE WINNER OF THE MOST SUCCESSFUL PRODUCT IDEA OF THE YEAR . . . IS . . ."

Chapter Five - Evaluating the New Product Idea

After you have gathered the necessary information and filled out most of the worksheets you are ready to evaluate the product for success potential. It is unlikely that you will have been able to find all of the information that you need but you should now have enough specific information about your target market and the competition so that any guessing you must do will be the educated kind.

QUESTION #1 - IS THE MARKET SIZE LARGE ENOUGH?

You should be able to estimate the total size of the market from sales figures for competing manufacturers, distribution sizes of periodicals serving your market and market data from media kits. The answer to the question of whether this is large enough will depend on how much profit you receive from the sale of each item, your market share for the product and your minimum requirements for profit.

Manufacturer's profit varies with the type of industry but 5% of the retail sale price is a good conservative estimate for most products. Market share is discussed under question #3.

As an illustration, let us assume that the size of your market is 3.5 million people. Also, assume that you can conservatively sell to 1% of this population each year and make a profit of 5% of the estimated sales price of $20. The math goes like this...

annual sales volume = 1% x 3.5 million = 35,000 units

annual sales = $20 x 35,000 = $700,000

annual profit = 5% x $700,000 = $35,000

If you are a small company or individual with small profit requirements and the product can be produced without having to purchase expensive molds and tooling, then $35,000 annual profit may be enough. If, however yours is a large company with large overhead expenses, $35,000 would probably not be equitable.

QUESTION #2 - CAN THE MARKET BE REACHED EASILY?

If your potential customers can be reached by the particular stores in which they shop, the organizations they belong to, or through the particular magazines they read, then your market can be targeted easily and advertising will be most cost effective. Rate your product a 5 if this is the case.

If your market cannot be characterized by any of the above criteria, then your advertising will necessarily be to the general population and therefore large and expensive.

QUESTION #3 -CAN YOU EXPECT A LARGE ENOUGH SHARE OF THE MARKET?

Market share will depend on how the customer views your product relative to the competition and how successful your efforts are in marketing and advertising. Generally, the product's market share will grow each year until another new competitive product is introduced and reduces the number of sales of your product. Typically, this cycle takes about 5 years. *Editor's Note: Many successful inventors continue trying to obsolete their invention by improving it over and over because they realize that if they do not improve it, others will. Doing this will help you keep your share of the market.*

If you write a questionaire for your potential target, include questions asking if they would purchase your product or your competition's. Include several price ranges for your product. Assuming your questionaire is statistically distributed amongst your target market, the percentage of positive responses can then be taken as your product's market share.

If you are unable to obtain customer opinions, attempt to make a conservative estimate of your product's market share by taking 10% of the smallest market share of your competitors. Another conservative estimate would be to assume that 1% of your target market will purchase your product each year. These numbers are only a little better than guesses since their calculation depends only on the size of your competitor's markets and exclude your product's placement among the competition.

QUESTION #4 - ARE THERE SEVERAL MARKETS FOR THE PRODUCT?

If there are several well-defined target markets for your product, the diversity of these markets will tend to stabilize sales fluctuations in any one of the markets (score = 5 or 4). A product with stable sales volume allows its manufacturer to reduce its costs through production planning techniques, and is always more desirable than a product having large fluctuations of sales.

QUESTION #5 - WILL THE INDUSTRY CONTINUE TO GROW?

Industries go through the same type of growth and decline cycle that new products do. Many new companies are formed each year to fulfill the large demand for products when an industry is in its growth stage (score = 5). As the number of companies in the industry grows, the number of products produced by the companies match the demand and the industry is mature (score = 4). As more companies are added to the industry, the industry reaches its saturation stage (score = 3). After an industry reaches saturation, profits tend to drop off as the supply is greater than the demand and companies must reduce prices to stay competitive (score = 2 or 1).

QUESTION #6 - IS THE MARKET STEADY?

If the market does not fluctuate with the general economy or seasonally, then score a 5. If the market is very sensitive to the economy or has severe fluctuations with season, it will score a 1.

QUESTION #7 - WILL THE PRODUCT REVOLUTIONIZE THE INDUSTRY?

If the new product is exactly the same as the competition, score a 1 here. If the product is the long awaited solution to a persistent problem, then score a 5. Score between these extremes according to the uniqueness and usefulness of the product.

QUESTION #8 - IS THE PRODUCT'S PRICE LOWER THAN THE COMPETITION?

You may not know the exact price of your product yet, however, it is not necessary at this point. Compare your concept to that of a competing product which has the same sales volume that you expect. Generally, your product will be more expensive than this competing product if it has more parts, better materials or is built better.

QUESTION #9 - IS THE PRODUCT'S PERFORMANCE BETTER THAN THAT OF THE COMPETITION?

Using the information from the Product Comparison and Potential Customers' Opinions from Appendix I, rate your product's expected performance and benefits as perceived by the customer.

QUESTION #10 - ARE THERE ESTABLISHED DISTRIBUTION CHANNELS THAT THE PRODUCT CAN USE?

Just as it is less expensive to reach your target market with advertising through magazines and organizations that already serve your market, it costs less to reach your target market with your product through distribution systems that are already in place to serve your market. Review the information of how the competing products are distributed to determine if your product can use these same channels or if you will need to establish your own.

QUESTION #11 - IS THE PRODUCT VERY DIFFERENT FROM THE COMPETITION?

Review the features and benefits of the competing products and how the potential customer rates these to determine if your product has unique properties which are perceived to be valuable to the customer and which can be used to develop your product's niche. For example, will the goal for your product's niche be high quality, cost effectiveness or speed?

QUESTION #12 - ARE THERE ONLY A FEW BIG COMPETITORS?

The existence of competitors who are large corporations can make entry into an industry difficult for a new comer. Large manufacturers are usually diverse enough to be able to temporarily lower the price of their product below that of competing products made by new competitors - at least until they drive the new company out of business.

QUESTION #13 - WILL THE DEMAND FOR THE PRODUCT LAST SEVERAL YEARS?

Compare the expected market life of your product with that of the competition. A product which is designed to work with another product already on the market but due to be updated will have a short life (score = 1). A product designed for a fad market will also score low. A product which is expected to be in demand long after the competition has become obsolete will score a 5.

QUESTION #14 - WILL THE PRODUCT BE ACCEPTED BY THE PUBLIC AND CONSUMER ORGANIZATIONS?

If the product is so unique that there are no competing products, then the product will likely not be accepted without first developing the market through customer education. This adds considerable cost to a new product. Another reason for a product not being readily accepted is "bad press" from consumer organizations about products in your particular industry. If either of these conditions are true, score a 1 here. In the absence of any resistance by the market to the new product, score a 5.

QUESTION #15 - IS THE PRODUCT TECHNICALLY FEASIBLE?

Obviously, you should not waste your time evaluating products which are scientifically impossible (e.g. perpetual motion) but that is not what is meant here by "technically feasible." A technically feasible product is one which is based on existing knowledge (score = 5). If more research and experimentation is required before the final product can be designed, then the product is not technically feasible at the present time and you should score a 1.

QUESTION #16 - CAN THE DESIGN PROBLEMS BE SOLVED IN A SHORT TIME?

Review the design problems that you have identified in Appendix I. If there are no problems, score a 5. If the problems are trivial score a 4. If the problems cannot be solved within six months, score a 1.

QUESTION #17 - DOES THE PRODUCT USE READILY AVAILABLE MATERIALS?

If the materials used in the manufacture of the product are available from many suppliers, then the cost of these materials will generally be competitive and relatively low (score = 5). Materials which are only available through a few specialized suppliers will be comparitively expensive which will drive up the cost of your product (score = 1).

QUESTION #18 - ARE ONLY SIMPLE MANUFACTURING AND ASSEMBLY OPERATIONS NEEDED?

Presently, in the United States, labor is the greatest cost of most manufacturing operations. If your product can be mass produced by automatic machinery, with little human involvement, then score a 5 here. If the product requires many hours of skilled labor to assemble, then score a 1.

QUESTION #19 - IS THE PRODUCT PATENTABLE?

There is more to answering this question than just whether you might be able to get a patent on the product - a good patent attorney can get you a patent on some aspect of any product. You need to consider if a patent can be obtained which will prevent competitors from using some unique feature(s) of your product. This feature should give the customer a benefit which cannot be easily obtained by any other means which your competitors can come up with. If your product does have this type of feature, then score a 5. Most new consumer products do not have a feature like this and competitors easily "design around" their patents.

QUESTION #20 - ARE THERE FEW PRODUCT REGULATIONS IN THIS INDUSTRY?

The United States Government regulates new products in certain industries. Before a new product can legally be sold within these markets, they must meet certain regulations and may have to be tested and certified. Regulated industries include Food & Drug, Medical Devices and Toys. To determine what federal regulations your product may have to comply with, visit your local Law Library and review the Code of Federal Regulations (CFR). In addition to federal regulations, there may be standards set for products in your particular industry. Common standards include those set by the American Society for Testing and Materials (ASTM), the National Electrical Code (NEC) and the National Fire Codes (NFC). The expense of complying to regulations increases the cost of bringing your product to market.

"GLASS JAW" PETE LEARNS THE HARD WAY THAT IT IS GOOD TO HAVE AN INSIDE EDGE

CHAPTER SIX - INTERPRETING THE RESULTS

Adding the scores you gave in response to each question, you will find the probability of success of the new product idea (according to this evaluation). To score well in this evaluation, a new product idea must serve a large, stable, well-defined and easily-reached market. The product must have unique features and benefits to place it in a good market position compared with competing products. There must be no unusual problems in the design and manufacture of the product. Finally, the product should be in an industry which is in its growth stage, with good public approval and having no costly regulations.

Your product most likely did not score 100. If it did, then you should consider having someone else evaluate your product for you. Before I recommend to continue with the development of a product, it must score at least 85 points. Depending on the risk you are willing to take with development money, your benchmark may be higher or lower.

If your product has scored high in this evaluation, congratulations! Only one in ten will. The information that you have collected in the evaluation will give you an "inside edge" in writing the product specification and designing the product to place it in the best competitive position. I recommend using professionals experienced in marketing, engineering and manufacturing products in your industry to properly develop your product. ***Editor's note: One can not be all things at once. Rarely, can anyone be successful wearing many hats, such as: inventor; manufacturer; marketeer; distributor; retailer; etc. If you are the inventor, then keep trying to make the product better. Use professionals to do the rest.*** **Good luck and never give up!**

Appendices A through I

APPENDIX A

COMPETITIVE PRODUCT WORKSHEET

PRODUCT NAME:	
MANUFACTURER: ADDRESS:	
DISTRIBUTION CHANNEL:	
ADVERTISING METHOD:	
RETAIL PRICE:	
MAIN FEATURE OR BENEFIT:	
ADDITIONAL FEATURE:	
ADDITIONAL FEATURE:	
ADDITIONAL FEATURE:	
CUSTOMER COMMENTS:	
CUSTOMER COMMENTS:	
CUSTOMER COMMENTS:	

NOTES AND ATTACHMENTS:

APPENDIX B

COMPETITIVE MANUFACTURER WORKSHEET

MANUFACTURER: ADDRESS:	
ANNUAL SALES:	
MARKET :	
SALES FROM MARKET :	
MARKET SHARE :	
METHODS OF ADVERTISEMENT:	
NICHE: WHAT IS COMPANY KNOWN FOR?	

NOTES AND ATTACHMENTS:

APPENDIX C

CATALOG COMPANIES WORKSHEET

CATALOG:	
MARKET:	
COMPANY NAME: ADDRESS:	
ADVERTISEMENT CONTACT:	
PHONE NUMBER:	

CATALOG:	
MARKET:	
COMPANY NAME: ADDRESS:	
ADVERTISEMENT CONTACT:	
PHONE NUMBER:	

CATALOG:	
MARKET:	
COMPANY NAME: ADDRESS:	
ADVERTISEMENT CONTACT:	
PHONE NUMBER:	

CATALOG:	
MARKET:	
COMPANY NAME: ADDRESS:	
ADVERTISEMENT CONTACT:	
PHONE NUMBER:	

APPENDIX D

PERIODICAL WORKSHEET

MAGAZINE:	
MARKET:	
MARKET SHARE:	
DISTRIBUTION SIZE:	
PUBLISHER: ADDRESS:	
ADVERTISEMENT CONTACT:	
PHONE:	

MAGAZINE:	
MARKET:	
MARKET SHARE:	
DISTRIBUTION SIZE:	
PUBLISHER: ADDRESS:	
ADVERTISEMENT CONTACT:	
PHONE:	

MAGAZINE:	
MARKET:	
MARKET SHARE:	
DISTRIBUTION SIZE:	
PUBLISHER: ADDRESS:	
ADVERTISEMENT CONTACT:	
PHONE:	

Appendix E

Patent and Trademark Depository Libraries

(Note: From USPTO website)

The following libraries, designated as Patent and Trademark Depository Libraries (PTDLs) receive current issues of U.S. Patents and maintain collections of earlier-issued patents as well as trademarks published for opposition. The scope of these collections varies from library to library, ranging from patents of only recent years to all or most of the patents issued since 1790 and trademarks published since 1872.

These patent and trademark collections, which are organized in number sequence, are available for use by the public free of charge. Each of the PTDLs, in addition, offers supplemental reference publications of the U.S. Patent Classification System, including the Manual of Classification, Index to the U.S. Patent Classification and Classification Definitions, and provides technical staff assistance in using such publications in gaining effective access to information contained in patents and trademarks. CASSIS (Classification and Search Support Information System) and other CD-ROM products for searching patent and trademark information are available at all PTDLs . Facilities for making paper copies of patents and trademarks from either microfilm or paper collections are generally provided for a fee.

Since there are variations in the scope of patent collections among the PTDLs and in their hours of service to the public, anyone contemplating use of the patents at a particular library is urged to contact that library, in advance, about its collection and hours in order to avert possible inconvenience.

STATE	LIBRARY	PHONE
Alabama	Auburn, Auburn University Libraries	(334) 844 - 1747
	Birmingham, Birmingham Public Library	(205) 226 - 3680
Alaska	Anchorage, Z J Loussac Public Library	(907) 562 - 7323
Arizona	Tempe, Noble Library, Arizona State University	(602) 965 - 7010
Arkansas	Little Rock, Arkansas State Library	(501) 682 - 2053
California	Los Angeles, Los Angeles Public Library	(213) 612 - 3273
	Sacramento, California State Library	(916) 654 - 0069
	San Diego, San Diego Public Library	(619) 236 - 5813
	Sunnyvale, Patent Information Clearinghouse	(408) 730 - 7290
	San Francisco, San Francisco Public Library	(415) 557 - 4488
Colorado	Denver, Denver Public Library	(303) 640 - 8847
Connctct.	New Haven, Science Park Library	(203) 786 - 5447
Delaware	Newark, University of Delaware Library	(302) 831 - 2965
D.C.	Washington, Howard University Libraries	(202) 806 - 7252
Florida	Fort Lauderdale, Broward County Main Library	(305) 357 - 7444
	Miami-Dade, Miami-Dade Public Library	(305) 375 - 2665

STATE	LIBRARY	PHONE
Florida	Orlando, University of Central Florida Libraries	(334) 844 - 1747
	Tampa, Tampa Campus Lib., Univ. of S. Florida	(813) 974 - 2726
Georgia	Atlanta, P Glbrt Mem Lib., Georgia Inst. of Tech.	(404) 894 - 4508
Hawaii	Honolulu, Hawaii State Public Library System	(808) 586 - 3477
Idaho	Moscow, University of Idaho Library	(208) 885 - 6235
Illinois	Chicago, Chicago Public Library	(312) 747 - 4450
	Springfield, Illinois State Library	(217) 782 - 5659
Indiana	Indianapolis, Indianapolis-Marion Co. Pub. Lib.	(317) 269 - 1741
	West Lafayette, Purdue University Libraries	(317) 494 - 2873
Iowa	Des Moines, State LIbrary of Iowa	(515) 281 - 4118
Kansas	Wichita, Ablah Library, Wichita State University	(316) 689 - 3155
Kentucky	Louisville, Louisville Free Public Library	(502) 561 - 8617
Louisiana	Baton Rouge, T H Mddltn Lib, Louisiana St Univ.	(504) 388 - 2570
Maine	Orono, R H Fogler Lib, University of Maine	(207) 581 - 1678
Maryland	College Park, Eng & Phys Sc. Lib, U of Maryland	(301) 405 - 9147
Mass.	Amherst, Phys Sc. Lib, University of Mass.	(413) 545 - 1370
	Boston, Boston Public Library	(617) 536 - 5400 X265
Michigan	Ann Arbor, Eng Transprtn Lib, Univ of Michigan	(313) 764 - 7494
	Big Rapids, A S Timme Lib., Ferris State Library	(616)592 - 3602
	Detroit, Detroit Public Library	(313) 833 - 1450
Minn.	Minneapolis, Minneapolis Public Lib & Info Cntr	(612) 372 - 6570

STATE	LIBRARY	PHONE
Mississippi	Jackson, Mississippi Library Commission	Not Operational
Missouri	Kansas City, Linda Hall Library	(816) 363 - 4600
	St. Louis, St. Louis Public Library	(314) 241 - 2288 X390
Montana	Butte, Montana College of Mnrl Sc & Tech Lib	(406) 496 - 4281
Nebraska	Lincoln, Eng Lib, Univ of Nebraska-Lincoln	(402) 472 - 3411
Nevada	Reno, University of Nevada-Reno Library	(702) 784 - 6579
N Hmpshr	Durham, Univ of New Hampshire Library	(603) 862 - 1777
N Jersey	Newark, Newark Public Library	(201) 733 - 7782
	Piscataway,Lib of Sc & Med, Rutgers Univ.	(201) 932 - 2895
N Mexico	Albuquerque,Univ of New Mexico Gen Lib.	(505) 277 - 4412
New York	Albany, New York State Library	(518) 473 - 4636
	Buffalo, Buffalo & Erie County Public Library	(716) 858 - 7101
	New York, New York Pub Lib, The Research Lib	(212) 714 - 8529
N Carolina	Raleigh, D H Hill Lib., N Carolina State Univ.	(919) 515 - 3280
N Dakota	Grand Forks, Chester Fritz Lib, U of N Dakota	(701) 777 - 4888
Ohio	Cincinnati, Cinti & Hamilton County, Pub Lib	(513) 369 - 6936
	Cleveland, Cleveland Public Library	(216) 623- 2870
	Columbus, Ohio State University Libraries	(614) 292 - 6175
	Toledo, Toledo/Lucas County Public Library	(419) 259 - 5212
Oklahoma	Stillwater, Oklahoma State University Library	(405) 744 - 7086
Oregon	Salem, Oregon State Library	(503) 378 - 4239

STATE	LIBRARY	PHONE
Penn.	Philadelphia, The Free Library of Philadelphia	(215) 686 - 5331
	Pittsburgh, Carnegie Library of Pittsburgh	(412) 622 - 3138
	University Park, Pattee Lib, Penn State Univ	(814) 865 - 4861
P.Rico	Puerto Rico, Mayaguez, Gen Lib, Univ of P.Rico	Not Operational
Rhode I.	Providence, Providence Public Library	(401) 455 - 8027
S Carolina	Clemson, Clemson University Libraries	(864) 656 - 5168
S Dakota	Rapid City, Dvrx Lib, S D Sch of Mines & Tech	(605) 394 - 6822
Tennessee	Memphis, Memphis & Shelby County Pub Lib	(901) 725 - 8877
	Nashville, Stevenson Sc Lib, Vanderbilt Univ	(615) 322 - 2775
Texas	Austin, McKinney Eng Lib, U of Texas at Austin	(512) 495 - 4500
	College Station, S C Evans Lib, Texas A&M Univ	(409) 845 - 2551
	Dallas, Dallas Public Library	(214) 670 - 1468
	Houston, The Fondren Library, Rice Univ	(713) 527 - 8101 X2587
Utah	Salt Lake City, Marriott Lib, University of Utah	(801) 581 - 8394
Virginia	Richmond, J B Cabell Lib, Virginia Comm Univ	(804) 828 - 1104
Washgton	Seattle, Eng Lib, University of Washington	(206) 543 - 0740
W Virginia	Morgantown, Evansdale Lib, W Virginia Univ	(304) 293 - 4510
Wisconsin	Madison, K F Wendt Eng Lib, U of Wis-Madison	(608) 262 - 6845
	Milwaukee, Milwaukee Public Library	(414) 278 - 3247
Wyoming	Casper, Natrona County Public Library	(307) 237 - 4935

INSTRUCTIONS ON HOW TO USE THE PATENT SEARCH WORKSHEET

1) Make at least ten copies each of the worksheets in appendix F and G.

2) Go to the Patent Depository Library and search through the Index to the US Patent Classification System for every word that describes or relates to the product that you are researching. Write each word that you find in the "SUBJECT" box on the Patent Search Worksheet. Write the class and subclass designations in the "CLASS/SUB" box.

3) Search through the Manual of Classification as discussed on page 27 for additional class and subclass numbers. Write each new classification number in a separate "CLASS/SUB" box on the Patent Search Worksheet. You will probably need several copies of the worksheet to list all of the classification numbers you will find.

4) Look up each classification number using CASSIS to review a list of titles of patents which are in your selected classification. Write down every patent which appears to be closely related to your product.

5) As you continue your search using the Patent Gazette and the Patents note if you have made copies of the abstracts or the patent itself.

APPENDIX F

PATENT SEARCH WORKSHEET

SUBJECT:	
CLASS/SUB:	

PATENT #	☑ COPY ABSTRACT?	☑ COPY PATENT?

SUBJECT:	
CLASS/SUB:	

PATENT #	☑ COPY ABSTRACT?	☑ COPY PATENT?

SUBJECT:	
CLASS/SUB:	

PATENT #	☑ COPY ABSTRACT?	☑ COPY PATENT?

SUBJECT:	
CLASS/SUB:	

PATENT #	☑ COPY ABSTRACT?	☑ COPY PATENT?

APPENDIX G

COMPETITIVE PATENT WORKSHEET

PATENT NUMBER:	
PATENT TITLE:	
INVENTOR NAME: ADDRESS: PHONE:	
ATTORNEY NAME: ADDRESS: PHONE:	
NOTES:	

PATENT NUMBER:	
PATENT TITLE:	
INVENTOR NAME: ADDRESS: PHONE:	
ATTORNEY NAME: ADDRESS: PHONE:	
NOTES:	

APPENDIX H

POPULAR INTERNET SEARCH ENGINES

(Note: Locations and Engines may have changed since this list was made.)

SEARCH ENGINE	LOCATION
YAHOO!	http://www.yahoo.com/search.html
LYCOS	http://www.lycos.com/
INFOSEEK	http://www.infoseek.com
GALAXY	http://www.einet.net/cgi-bin/wais-text-multi?
WEBCRAWLER	http://webcrawler.com/WebCrawler/WebQuery.html
DEJANEWS	http://www.dejanews.com
SNOOPIE	http://www.snoopie.com
SAVVY SEARCH	http://rampal.cs.colostate.edu:2000/
CUSI	http://www.qdeck.com/cusi.html
OPEN MARKET	http://www.directory.net/
other sites	http://www.yahoo.com/Reference/Searching_the_Web/

APPENDIX I

MARKET WORKSHEET

CUSTOMER INFORMATION

LEAD USER INFORMATION

APPENDIX I

MARKET WORKSHEET

PROBLEM

HOW THE PRODUCT SOLVES THE PROBLEM

APPENDIX I

MARKET WORKSHEET

VALUE OF THE PRODUCT

PRODUCT COMPARISON TO THE COMPETITION

APPENDIX I

MARKET WORKSHEET

POTENTIAL CUSTOMERS' OPINIONS

SIZE OF THE MARKET(S)

APPENDIX I

MARKET WORKSHEET

DESIGN PROBLEMS

MARKET TRENDS

RAPID PRODUCT DEVELOPMENT PRESS

ORDERING INFORMATION

DID YOU BORROW THIS BOOK?

DO YOU WANT A COPY FOR YOURSELF?

CAN'T FIND IT IN THE BOOKSTORE?

YOU CAN GET ADDITIONAL COPIES OF THIS BOOK DIRECTLY FROM THE PUBLISHER BY SENDING A CHECK OR MONEY ORDER IN THE AMOUNT OF $19.95 FOR EACH COPY REQUESTED. ORDER FIVE OR MORE COPIES AND TAKE A 20% DISCOUNT! GREAT FOR INVENTOR'S GROUPS!

SEND YOUR REQUEST AND REMITTANCE TO:

"INVENTOR'S GUIDE TO IDENTIFYING PROFITABLE IDEAS"

RAPID PRODUCT DEVELOPMENT *Press*

P.O. BOX 655

AMELIA, OHIO 45102-0655

ABOUT THE AUTHOR:

Jon W. Mooney has an Aerospace Engineering degree from the University of Cincinnati. During the Cold War, Jon worked on diverse military and government projects, including SkyLab, the Space Shuttle, Patriot Missile and classified programs. After the end of the Cold War, Jon worked to transfer military technology into commercial products. During this time, he gained substantial knowledge of the requirements for a commercially successful product. "The secret to successful product development is to spend the least resources to identify and throw out the majority of ideas which have no chance for success." Jon shares his inexpensive techniques through consulting, seminars and now through this book.

ABOUT THE EDITOR:

John L. Janning is one of Dayton, Ohio's best known and most prolific inventors. Presently he has 34 U.S. Patents and over 200 worldwide. Formerly an engineer with NCR, John now operates his own lab, where he develops and markets his inventions and performs research and development for hire.